Dale and I like to fish.

Dale is my sister.

She and I go to the pond.

I have two hooks in a cork.

We have two poles with us.

We are set to go.

I cast my line into the pond.
Dale gets her line set.

In no time, my line jerks.
I have a bite!

I grit my teeth.
I tug the pole back.

I see the fish when I reel it in.

Is it big?

Yes, it is big!

Dale grabs the net.
She scoops up the fish.

This is one big fish!
It will be a fine lunch.